AVIATION

Aviation

Early Years of Flight
Prior to 1934

Rosary Spera Abrams

Pendleton Artists
California

Aviation, Early Years of Flight Prior to 1934, Copyright © 2015 by Linda Pendleton, All Rights Reserved.

ISBN: 9798660089558
Imprint: Independently published

No part of this book may be reproduced or transmitted in any form or by any means, graphic, digital, electronic, or print, including photocopying, recording, taping, or by any storage retrieval system without the written permission of copyright owner.

Cover Design by Judy Bullard, Cover Concept by Linda Pendleton.
Cover Airplane, De Havilland DH-4, National Museum of the United States Air Force. (U.S. Air Force photo).

Pendleton Artists
California

www.lindapendleton.com
First Kindle Edition, 2015.
First Print Edition, 2020.

To my brother, Nick Spera, I dedicate this project.
~Rosary Spera, 1933.

For our family; in honor and loving memory of Rosary Spera Abrams.
~Linda Pendleton, 2015.

Man must rise above the Earth—to the top of the atmosphere and beyond—for only thus will be fully understand the world in which he lives.
~Socrates

Rosary Spera, 1937

Foreword

The blue sky above has always inspired men and women to devise ways to spread their *wings* and move above the earth in a desire to soar like an eagle. Over the years, for some, the idea to achieve and master flight became a dream, a challenge, and even an obsession. Thank goodness for that, as we look back at the early history of aeronautics and realize the magnitude of what has been accomplished over time—decades of time.

The science of flight, the aerodynamics, design, workmanship, and operation of a craft, resulted from creative, visionary, skilled, and daring minds. Those who chose to pilot the inventions, so often lived on the edge of danger as a balloon, dirigible, or plane rose from the ground and upward, with hopes of staying in the air in a controlled flight until time to return safely to the ground.

The late 18th century, the 19th century, and the first three decades of the 20th century were exciting times of ongoing advancements as new crafts were designed, and new flight records set, while crowds of enthusiastic spectators enjoyed air shows featuring the latest in aviation, aerial stunts, and internationally recognized pilots. America became captivated by what was happening in the cerulean blue skies above.

Rosary Spera was one who had an interest in the advances of aviation, and in 1933, while a student, she wrote this book.

My mother, Rosary, dedicated this book to her older brother, who, with his interest in aviation, went on to have a lifelong career in aviation.

It was a time of an exciting momentum that ultimately would change the world.

Rosary Spera Abrams died in 2012 at the age of 93 years. I believe my mother would be honored to know her 1933 research on flight will now be shared with others who were not alive in those earlier years to experience the exhilaration of the explosion of inventions and historical advancements.

I also wonder what those of the earliest days of aviation would now think of the incredible technology and advances that have been made in the science of flight. Could they have even imagined?

~Linda Pendleton
California, 2015

Pioneers of Flight

232 B.C. to 1911 A.D.

The foundation of man's flight was laid when ancient Greek mathematician, physicist, and inventor, Archimedes of Syracuse, Sicily, propounded the law governing the flotation of bodies in liquids and gases. The principle indicates that the upward buoyant force that is exerted on a body immersed in a fluid, whether fully or partially submerged, is equal to the weight of the fluid that the body displaces. But it was not until 1783 that man first rose in a balloon, although the intervening centuries gave birth to many legends of daring souls who attempted to fly.

Archimedes of Syracuse

Nearly 70 years passed before the balloon became steerable, and another century was turned before the Wright brothers, Orville and Wilbur, inspired by the glider experiments of German-Prussian, Otto Lilienthal, and others, succeeded in ascending in a motor driven heavier-than-air machine on December 17, 1903. Then, came a series of experimental flights and each year saw the fledgling wings increase in power.

AVIATION

Louis Charles Blériot, 1909

In 1909, French engineer and inventor, Louis Charles Blériot made history with the first flight of a heavier-than-air plane, across the English Channel. He had successfully manufactured automobile headlamps and used some of the resulting funds to build his airplane. His flight took 36 minutes and 30 seconds, landing not far from Dover Castle.

Glenn Curtis, *June Bug, 1908*

AVIATION

Glenn Curtis

The first long-distance flight between two major cities in the United States was made on May 29, 1910, by aviator and aircraft builder, Glenn Curtiss, who flew from Albany, New York to New York City. The 137 mile flight resulted in Curtiss winning a $10,000 prize by publisher, Joseph Pulitzer and was awarded the Scientific American trophy.

Ralph Pulitzer, eldest son of Joseph Pulitzer, who took over the newspapers following his father's death in 1911, founded the Pulitzer Trophy Races in 1920, also known as the National Air Races. A series of cross-country air races resulted in advances in the science of aviation, and speed and reliability of aircraft and aircraft engines grew rapidly.

Calbraith Perry Rodgers

On September 17, 1911, Calbraith Perry Rodgers began the first transcontinental flight across the United States, from Long Island, New York landing in Pasadena, California on November 5,

2011. He flew the Vin Flyer, an early Wright Brothers Model EX biplane.

On April 3, 1912, Rodgers died while making an exhibition flight over Long Beach, California, when he flew into a flock of birds, causing the plane to crash into the ocean.

Calbraith Perry Rodgers, 1911
Vin Biplane

AVIATION

It was not until 1923 that the transcendental nonstop flight was made in one hop. U.S. Air Service Lieutenants John A. Macready and Oakley G. Kelly in a Fokker T-2 airplane took off from Roosevelt Field, Long Island, New York, on May 2, 1923, and almost 27 hours later, the single-engine plane landed at San Diego, Calif., having flown a distance of 2,520 miles. For their successful nonstop flight across the United States, Macready and Kelly were awarded the Mackay Trophy for 1923.

Fokker T-2

Rosary Abrams

War Birds and Record Makers

The years between 1911 and 1923, saw the development of aviation from a precarious plaything to an established industry. World WarI did much to bring it to a quick maturity, with the fate of nations, perhaps, depending on aerial superiority. The years of conflict brought forth many illustrious aces of all nations, who demonstrated what could be done with a plane. The war time pace was not slackened with the coming of peace, but was constantly accelerated. Achievements became more and more ambitious. Even "stunting" was turned to scientific account.

Achieving the Spectacular

Flights over the oceans and over the continents are only one phase of aviation's achievements. In many other ways, new records are being established. Planes are kept flying for days and weeks without landings. Others pierce even

AVIATION

higher into the rarified upper atmosphere where pilots must breathe oxygen to keep going. Speed records of all sorts are broken almost daily, bringing towns and even nations hours nearer each other. Other pilots demonstrated the reliability and utility of planes in other ways—blazing the trail for commercial aviation of next year and the year after next. The speed and altitude and endurance records of today indicate the routine pace of tomorrow.

The Influence of the War

World War I
1915 - 1919

There is not the slightest doubt that the World War did more for aviation during the period it ran its course, than might have been done during ten times that length of time had peace continued. It moved aviation at least half a century ahead. Recognizing the potentialities of airplanes for offense and defense, world powers spent money for airplane development liberally. Billions were made available where only hundreds of thousands might otherwise have been obtained. There was, to be sure, much waste of time, men and money, but out of these expenditures came the airplane as we know it today.

The post-bellum period produced conditions to which flying had some difficulties in adjusting itself. The airplane occupied a unique and not altogether advantageous position in the public mind. Its peacetime value was underestimated because it wartime value had been so thoroughly underscored. It slipped into its old position as a vehicle for thrills. Stunt flying once more came into vogue. The public was impressed by the airplane but leery of it. It caused too many deaths. Finally, thrill flying lost both flavour and favour. Air routes were established. Commercial flying concerns began doing business. They expanded and prospered.

Then came the Federal law of July 2, 1926, the Air Commerce Act, which gave the country a permanent policy for the development of military and commercial aviation. This landmark legislation charged the Secretary of Commerce with fostering air commerce, issuing and enforcing air traffic rules, certifying aircraft, establishing airways, licensing pilots, and operating and maintaining aids to air navigation. William P. MacCracken Jr. became the first director of the new Aeronautics Branch in the Department of Commerce which assumed primary responsibility for aviation oversight.

Finally—in 1927—the air-mindedness of America became thoroughly harnessed and directed, due to the epoch-making flights of Lindbergh and others.

There are, so far, three great years in aviation history: 1903 - 1909 - 1927.

Of these, 1903 will always be the greatest, and, because of this, Wilbur and Orville Wright will always stand foremost among men who have contributed gloriously to aeronautic progress.

Orville and Wilbur Wright

AVIATION

Spanning the Seas

1919 - 1931

A United States Navy plane made the first flight across the Atlantic in 1919 from North America to Europe with Lieutenant Commander Albert Read and his crew of five on the Navy seaplane NC-4. Reads crew consisted of Lieutenant Elmer Stone U.S. Coast Guard, as first pilot and Lieutenant W. K. Hinton as co-pilot. Other crew members of NC-4 were Ensign H.C. Rodd, radio operator; Lt. James Breeze, engineer: and Chief Machinists Mate E.C. Rhoads, relief engineer. The Aircraft Commanders were navigators and operated from the bow of the aircraft.

NC-4

AVIATION

They arrived in the Azores on May 27, 1919, after a harrowing flight from Newfoundland during which they encountered violent storms and disorientation and came close to crashing several times.

The crew of the NC-4 at Lisbon: From the left Chief Machinist's Mate (Air) E.S. Rhoads, USN, Engineer; Lt. W.K. Hinton, USNRF, Pilot; Lt. J.L. Breese, USNRF, Engineer; Lt. E.F. Stone, USCG, Pilot; LCdr. A.C. Read, USN, Commanding Officer and Navigator; Ens. H. C. Rodd, USNRF, Radio Operator does not appear in the picture.

The same year not only saw the first non-stop flight over that ocean, but the first dirigible crossing.

The first round trip across the Atlantic by airship began on July 2, 1919. The British Airship R35 flew from Scotland to Mineola, New York, a flight of 108 hours and 12 minutes, and a return home trip to Europe, arriving in Pulham, England on July 13, 1919.

Then there were several years of quiet until 1924, when two U.S. Army planes flew over the North Pole.

AVIATION

Charles Lindbergh and *Spirit of St. Louis*.
Library of Congress

Charles Lindbergh, a U.S. Air Mail pilot and a U.S. Army Air Corp Reserve officer, rose to prominence with his solo nonstop flight on the single-seat, single engine monoplane, *Spirit of St.*

Louis, from New York to Paris on May 20-21, 1927. Lindbergh was awarded the Medal of Honor for the historic event.

The Lindbergh flight gave impetus to a number of trans-oceanic flights, including Charles Edward Kingsford-Smith's flight from California to Australia, and the first successful westward crossing of the Atlantic by plane.

Naval officer Richard Byrd flew over the South Pole in November 1929. In 1930, two Frenchmen flew from Paris to New York; while 1931 saw a fleet of fourteen Italian planes fly the South Atlantic.

AVIATION

Chronological Review of Aviation History

232B.C.....Archimedes propounded law governing the flotation of bodies in liquids and gases.

1766....Properties of "inflammable air" or hydrogen, demonstrated by Henry Cavendish in England.

1783....August 27, Hydrogen balloon constructed at Paris, by Professor Charles, the physicist, flew 15 miles.

1784.....July 17, First balloon ascent in America made by Peter Carnes at Baltimore, Maryland, manned by 13 year old Edward Warren.

1794....The French revolutionary armies used a balloon for reconnaissances in their campaigns against the Austrians. The size of the balloon, however, and the difficulties of transportation and inflation, limited their usefulness as engines of war.

1821....August 19, Charles Green made the first ascent in a balloon filled with coal gas at the coronation of King George 1V. Since then, with exceptions, hydrogen gas has been used only for dirigibles.

1836....November 7, Charles Green, Robert Holland, and Monck Mason, flew from London to Weilburg, Germany, 500 miles away, in 18 hours.

1850....A model airship, shaped like a torpedo and driven by a screw propeller, was exhibited and flown at Paris, by the builders, Jullien and Jules Henri Gifford, of France.

1870....Autumn, Numerous escapes by balloon were made from Paris which was under siege. On November 24, two aeronauts leaving there were caught in a gale which landed them 13 hours later in Norway, 600 miles away.

1884....August 9, Captains Renard and Krebs, of the French Army, experimented with an eclectically propelled airship, La France, making for the first time, a completed circle in the air.

1897....May, Augusti Andrée of Sweden and two companions in a special constructed balloon took off from Dansk Gatt, northwest of Spitsbergen, in an attempt to fly to the North Pole. They never returned.

AVIATION

1900....Orville and Wilbur Wright experimented with gliders.

1903....December 17, The first flight in a heavier than air, motor-driven plane was made by Wilber and Orville Wright, United States, at Kitty Hawk, North Carolina.

1905...October 5, The Wright Brothers flew for a distance of about 25 miles in 38 minutes.

1911....The first airmail in the United States was carried from the Nassau Boulevard Airdrome, Long Island, to Mineola, Long Island.

1912....Airplanes were used in the Balkan Wars.

1914....The event of the World War caused airplanes to be used for the first time on a large scale.

1918....May 15, The first regular airmail service in the world was established between New York and Washington D.C.

1920....February 14 to May 31, Lieutenants Guido Masiers and Arturo Ferrari, Italy, flew from Rome, Italy to Tokyo, Japan, a distance of about 12,000 miles round trip, in an Italian Ansaldo SVA reconnaissance biplane.

1922....September 5-6, Lieutenant J. H. Doolittle, United States, flew from Jacksonville, Florida, to San Diego, California in 21 hours, 18 minutes, making one stop.

1924....April 6 to Sept. 28, United States aviators, in four Douglass Transport planes, left Seattle, Washington on round-the-world flight by the way of Alaska, Japan, India, Persia, Iraq, Turkey, Austria, England, Greenland, and Newfoundland. Two complete the distance of 27,553 miles safely. Actual flying time, 371 hours, 11 minutes, over a period of 175 days.

1925...April 21 to November 9, Commander Francisco de Pinedo, Italy, flew from Roma to Japan and return, by way of Melbourne, Australia, 35,000 miles.

1925....November 16 to March, 1926, A. J. Cobham, England, flew from London, England, to Cape Town, Union of South Africa, 8,000 miles.

1926....May 9, Commander R. C. Byrd and Floyd Bennett flew from Kings Bay, Svalbard (Spitsbergen) to North Pole. Returned in 15 hours.

1926....December 21 to May 2, 1927, U.S. Army airplanes made a good-will flight around the

AVIATION

Latin American countries, covering a distance of about 20,000 miles.

1926....May 20-21, Col. Charles A. Lindbergh, United States, made the first non-stop flight from the United States to Europe by flying from New York to Paris, a distance of 3,610 miles in thirty-three and one-half hours.

1927....June 4-5, Clarence D. Chamberlin and Charles Levine, United States, made a non-stop flight from New York to Eisleben, Germany, 3,905 miles, breaking Lindbergh's distance of two weeks earlier.

1927...June 28-29, Lieutenants L. J. Maitland and A.J. Hegenberger, U.S. Army, made a non-stop flight from Oakland, California to Honolulu, Hawaii, a distance of 2,400 miles in 26 hours.

1928....May 31, June 9, the "Southern Cross" a Fokker F. VII/3m trimotor monoplane flown by Captain C. Kingsford-Smith and C. Ulm, Australia, and H.W. Lyon, and J. Warner, United States, flew from Oakland, California to Sydney, Australia, stopping only at the Hawaiian Island, Fiji Island, and Brisbane, Australia, 7,200 miles.

1928....June 17-18, Miss Amelia Earhart was the first woman to fly across the Atlantic (as a passenger) when the "Friendship" piloted by

Wilmer Stultz and Louis Gordon, successfully negotiated a flight from Trepassey Bay, Newfoundland, to Burry Port, Wales. Earhart kept the flight log.

1928....October 11-15, The Graf Zeppelin, German dirigible commanded by Dr. Hugo Echener, made the first commercial transatlantic flight from Friedrichshafen, Germany, to Lakehurst, New Jersey, carrying 20 passengers and a crew of 40. The flight was 111 hours and 44 minutes. The return trip also was successful.

1929....May 9, World altitude record for airplanes, 39,140 feet, made by Lieutenant Apollo Soucek, United States Navy, at Anacostia, Washington, D.C.

1929....May 25, Curtiss Marine Trophy Race for seaplanes and flying boats held at Anacostia, D.C., won by Lieutenant W. G. Tomlinson, U.S. Navy in an XF7C-1 with a speed of 162.52 m.p.h.

1929....June 27-29, Captain Frank M. Hawks established three trans-continental records by flying from New York to Los Angeles in 19 hours, 10 minutes; returning to New York in 17 hours, 38 minutes; and making the round trip in 44 hours, 4 minutes.

AVIATION

1929....August 20, The first all-metal dirigible in the world, built for the U.S. Navy by the Detroit Aircraft Corp., made a successful trial flight at Detroit, Michigan.

1930....March 30-April 6, Captain Hawks in a glider, towed by plane from San Diego, California to New York.

1930....May 24, Amy Johnson, England, finished 9,900 miles flight from London to Australia, becoming the first woman to make a solo flight from England to Australia.

1930....August 13, Captain Hawks flew from Los Angeles to New York in 12 hours, 25 minutes.

1931....May 30-June 4, The DO-X, giant German plane, flew from Bolama, Africa to Natal, Brazil.

1931....June 23-July –Wiley Post, U. S., pilot, and Harold Gatty, Australia, navigator, circled the globe in 8 days, 15 hours, 51 minutes, a record-breaking 15,474-mile flight in the "Winnie May," a Lockheed Vega. Their flight broke the previous record for fastest around-the-world flight held by the German Zeppelin.

Lockheed Vega

Zeppelin

AVIATION

Useful Information of Aviation

The first flights in a heavier-than-air machine was made with gliders by Otto Lilienthal, a German experimenter, and by Octave Chanute, an American architect, who make a careful study of air reactions under both flat and cambered panes. These flights were made in gliders without power.

Otto Lilienthal

Otto Lilienthal, (1848-1896) was a German pioneer of aviation and an inventor. He was known to be the first person to make well-documented gliding flights.

In 1889, he published in German, his book, *Birdflight as the Basis of Aviation.* In 1911, it was translated into English.

His gliders were carefully designed to distribute weight evenly to assure a safe flight. Shifting of the body, changing the center of gravity, gave Lilienthal control of the craft.

He gained world wide recognition as news of his short airborne flights spread, and photographs appeared in popular publications.

Lilienthal Glider

Lilienthal soon became known as the "father of flight," having controlled a heavier than air craft in sustained flight. He made about 2,000 flights in gliders of his own design, beginning in 1891 with his first successful glider design, *Derwitzer*. He held 23 patents of various inventions.

On August 9, 1896, Otto Lilienthal died from injuries suffered when his glider stalled, and unable to regain control, the craft fell approximately 50 feet. It was his fourth flight of the day, and his last.

The Wright Brothers credit Lilienthal as a major influence in their decision to pursue manned flight. Wilbur Wright stated, "Of all the men who attacked the flying problem in the 19th century, Otto Lilienthal was easily the most important." That being said, the Wright Brothers abandoned his aeronautical date after gliding for awhile and began using their wind tunnel data.

Octave Chanute

Octave Chanute was born in 1832 in France and died in Chicago, Illinois in 1910. He was a railroad civil engineer and aviation pioneer. During his career he designed and constructed bridges, including the Hannibal Bridge, competed in 1869, and the first bridge to cross the Missouri River in Kansas City, Missouri. He designed and built the Chicago and the Kansas

City Stockyards. He also devised a preservation treatment for railroad ties.

He first became interested in aviation in 1856 after watching a balloon flight. Upon retiring from his railroad engineering and consulting career, he devoted his time to the new science of aviation. A series of his articles were published in 1891-93, and in 1894, republished in the book, *Progress in Flying Machines.*

Chanute's aviation interest led him to design gliders after testing the glider design of Lilienthal. He went on to design hang gliders as well as gliders, with William Avery and Augustus M. Herring.

Chanute introduced the "strut-wire" wing structure used in bi-planes. His design was based on the Pratt truss, invented in 1844 by Thomas and Caleb Pratt, and used in railroad bridges.

He publicized the work of the Wright brothers and graciously shared his aviation knowledge with whoever was interested. But differences came up between him and the Wright brothers as they were withholding of their knowledge, and Chanute felt differently.

In 1872, Chanute, Kansas was named in his honor.

Louis Blériot

Louis Blériot was born in 1872, in Cambrai, France. He is an inventor and electrical engineer. He developed the first practical headlamps for automobiles, and in 1897, opened a successful business in Paris supplying two of France's automobile manufactures with lamps.

After developing several planes, he created the world's first successful powered monoplane.

On July 25, 1909, Blériot successfully piloted his Blériot X1, a monoplane powered by a 25-horsepower engine, from Calais, France across the English Channel to Dover, England. This won him a prize of £1,000 offered by the London Daily Mail Newspaper and resulted in international recognition as a leading pilot and aircraft manufacturer.

The monoplane of his design and construction was powered by a small three cylinder motorcycle type air-cooled engine. He flew the twenty-one miles between Calais and Dover in twenty-three minutes, an average speed of fifty-three miles per hour. The aviator had perfect control over his machine. After the flight, he even circled over the English shore until he found a suitable landing place.

Blériot XI Monoplane, 1911
U.S. Air Force Photo

Harriet Quimby

Harriet Quimby, (1875-1912), aviator, was also a journalist, photographer, actress, and one of the first female screenwriters.

On August 1, 1911, Harriet became America's first licensed woman pilot. On September 4, 1911, she piloted a Moisant built monoplane in the first night flight recorded by a woman, at a Richmond, California air show.

In 1912 she flew a Blériot XI from England to France across the English Channel. A few months after her successful flight across the Channel, she died during a flight at the Third Annual Boston Aviation Meet, near Boston, when the Blériot went into a nose-dive. Harriet Quimby and her passenger fell from the plane to their deaths in Dorchester Bay.

Rosary Abrams

Harriet Quimby
Library of Congress

AVIATION

Aeronautics

The study of aeronautics includes the study of the air, the weather, all types of balloons and flying machines, their parts, construction, operation, maintenance and repair, flying fields, and in fact, everything having to do with aerial navigation.

Aviation is really the practical use of aero dynamical flying machines. Common usage has made the word applicable to all types of aircraft.

Aerostation is the practical use of all kinds of flying craft that depend mainly on the lighting power of light gases for sustentation.

The term "aircraft" includes all types of balloons and airplanes.

Dirigibles are called ""blimps" because this is a contraction of an English designation for non-rigid airships which were known as "Class B-limp," to differentiate them from the rigid types. This colloquialism can only be applied to non-rigid balloons, though some people, especially newspaper writers, refer to all dirigibles as "blimps" erroneously.

The weather that is most unfavorable to dirigibles is sleet storms, because the precipitation freezes on the bag and the coating of ice on all parts greatly increase the weight.

Ordinary rain or snow does not materially influence navigation.

The "static" lift in a dirigible is the lifting power obtained by the gas being lighter than the air it displaces. The amount of static lift depends upon the gas employed and its purity as well as the capacity of the gas container.

The "dynamic lift" is the lift due to air pressure under the hull as it is forced through the air by the propulsive power of the engines.

The only effect rain has on a dirigible is it cools the gas and decreases its volume and the effective lift.

The wings of an airplane do all the lifting, because they have a large surface against which the air can lift.

The aerial screw or propeller, which is turned by a suitable gasoline engine, drives a plane through the air.

A cambered airplane wing is known as an "aerofoil."

An airplane will land at about half its maximum flying speed.

The cruising radius is the air distance the plane can cover without refueling.

A biplane has an extra set of wings, wing and interplane struts and bracing wires.

AVIATION

Mono-Spar Plane

Airplanes with folding wings have been built for years, but a radically new style of folding craft recently made its appearance at the Croydon Airdrome near London, England. Known as the "mono spar plane," it is expected to solve the problem of reducing weight, and increasing cargo and passenger capacity. The wings swing into lying position upon flying at the leading edge, while auxiliary sections swing downward from the fuselage to complete the unbroken wing lines. Two motors operate the twin propellers at the front of the craft.

"Skyspeaking"

Voice from plane heard for miles. Loud speakers often have been fitted to airplanes, so that a flyer may address a message to a crowd on the ground. So huge are the resonators in a new German plane designed for "skyspeaking," that they had to be built into the wings and aerodynamically shaped. It is said that, speaking from this plane in a low voice, one may be heard for miles.

The Dornier DO-X

The wings of this huge bird, span 157 feet. From tip to tail it is 137 feet. It has three decks. In the lowest deck are fuel, freight, tools, and food when needed. In the highest deck, the pilots, navigators, engineers, radiomen, and crew carry on their duties. The middle deck is for passengers, with plenty of room. The deck for the passengers has sleeping quarters, as well as sitting and dining rooms. There is room for 100 passengers and a crew of 12. But in the mighty test of a few weeks ago, this number was swelled by 57 more, 167 in all.

Twelve great engines of 500 horsepower each, provide the power for this liner of the air. With all 12 engines roaring at once, this great bird can dash 150 miles in one hour, though 115 miles per hour as the cruising speed. The entire ship weights 54,872 pounds. People can crowd into it until, with fuel for a flight a burden of 113,300 pounds may be lifted into the air. This is approximately 55 1/2 tons.

Two additional Dorniers have been ordered. We shall watch with interest for the first visit, for it seems that all great things come to America sooner or later.

AVIATION

May 30-June 4, 1931, the semi-cantilever giant German monoplane, DO-X, flew from Bolama, Africa to Natal, Brazil.

The first visit to America occurred on August 27, 1931, when the DO-X arrived in New York.

DO-X

Rosary Abrams

Tunnel Pierces New Airship

June, 1933. A thirty-three foot model of a new style of dirigible was demonstrated at Van Nuys, California, the other day. An open tunnel pierces the entire length of the gas bag from bow to stern. The inventor and patent holder, Thad Rose, believes that this will reduce wind resistance and increase the speed of the ship in flight. Two small electric motors ran the model airship's tiny propellers in the successful trial flights.

It was reported in a September 18, 1930 edition of the *Decatur Evening Herald* newspaper that observers who had seen the model believe that this type of dirigible promises to revolutionize the science of aviation.

AVIATION

Plane Takes Off From Car Roof

August 20, 1932. Catapulting an airplane into the air from the top of a speeding automobile was a feat carried out successfully, the other day, at Los Angeles, California. A special platform was built on the roof of a standard sedan. When the airplane was in place and the pilot ready at the controls, the car started down the Metropolitan Airport field at forty-five miles an hour. The pilot O. C. Le Boutillier found that he could launch his machine from this novel catapult three-hundred feet from the start. This method may be used to launch planes in small places.

Oliver Collin Le Boutillier was born in New Jersey in 1894. He trained at the Wright School of Aviation. In 1916, he joined the British Royal Naval Air Service, and in 1918, was one of four Americans who participated in the air fight that brought down the ace German pilot, Manfred von Richthofen, known as *der Rote Baron* or Red Baron.

After returning to America following World War I, he became a skywriter, and in 1928 was pilot for socialite, Mabel Bolls in the failed attempt for her to become the first woman to cross the Atlantic.

Oliver Collin Le Boutillier, 1918

AVIATION

Newspaper Talk
Los Angeles Record Newspaper

A One Man Around the World Flight

Dateline: Omsk, Siberia, June 6, 1933
American navigator, James J. Mattern of Texas, flying alone around the world, decided to spend the night in Omsk before continuing his journey across Siberia.

Mattern made the 1300 hundred mile flight from Moscow in 11 hours and 35 minutes, falling short of the Wiley Post and Harold Gatty time for the same distance by 3 hours and 30 minutes. The margin of his lead over the Post and Getty record is reduced to 1 hour and 35 minutes.

He was sighted only once on the long flight across Russia, to Omsk, Siberia just over the Ural Mountains and near the border between Asia and Europe. He was sighted in the night sky by inhabitants of Kazan, traveling high and fast, 3 hours and 5 minutes after takeoff from Moscow.

Mattern arrived in Moscow after a fast flight from Oslo, Norway. He had intended to take off again in two hours but minor motor repairs held him up for 9 hours and 25 minutes.

Post and Gatty lost 11 hours and 30 minutes in Moscow.

After landing in Omsk, Mattern retired for two hours of sleep. A check of his engine revealed the necessity for repairs, and Soviet officials lighted ground flares. Giving personal attention to every detail, he labored leisurely, assisted by Soviet mechanics. With repairs completed, he spoke for a few minutes with the United Press correspondent.

He is quoted as saying, "I gave my machine a much needed overhauling which I could not have done so well in the wilds of Siberia. I found a big piece of felt in the gasoline filter, which reduced the pressure. That explains some bad jolts I had crossing the ocean. Finding it now probably saves my neck."

Asked by the correspondent if he were trying to beat the Post and Gatty record, he said, "If I can lop off a couple of days from their record, so much the better, but that would be gravy, as I'm trying to establish a record for a one man around the world flight."

AVIATION

U.S. World Flier Wings Over Siberia

Dateline: Novosibirsk, Siberia, June 7, 1933
Jimmy Mattern, American Aviator, was sighted at 1:10 a. m. (Greenwich Meantime) today, enroute to Chita. Mattern was flying so high he was barely visible. Novosibirsk is approximately 400 miles east of Omsk.

Dateline: Moscow, June 7, 1933
Soviet officials felt today it was likely Jimmy Mattern had landed in Chita, Siberia. News of his arrival may be long delayed as much as 10 to 20 hours.

News of his plane being sighted over Novosibirsk was telephoned to the United Press bureau here. The telephone line ends at Novosibirsk. Officials at the airdrome there told the United Press that weather conditions were good and there was every reason to believe the flight was proceeding smoothly.

Flier Lands in Siberia

Dateline: Moscow, June 9, 1933
Jimmy Mattern took off from the airport at Krasnoyarsk, Siberia at 11:55 a.m. today, Moscow time after a five-hour stop to refuel and repairs on his plane. When leaving Krasnoyarsk,

Mattern was uncertain where he would stop next; Chita or Irkutsk. After taking off and heading eastward, minutes later he circled the air field and made a three-point landing. After that, he was delayed for two days for repairs.

Lost in the Arctic Tundra
Jimmy Mattern Missing

Four nearly four weeks searches were on to find James Mattern. About the time everyone had assumed the worst, a telegram was received which stated, "Safe at Anadyr, Siberia. Jimmie Mattern."

Facing a frozen oil line, Mattern was forced to belly land his plane in the tundra near the Anadyr River area. It was a fight for survival but finally he was found by Eskimos and taken to an outpost above the Arctic Circle at Anadyr. Not long after, Soviet aviator, Sigizmund A. Levanevsky flew to Anadyr to fly Mattern from Anadyr to Nome, Alaska.

AVIATION

Mrs. Franklin Roosevelt to be in Los Angeles Tonight
First Lady and Son Ready to Take Off at Tucson

Dateline: June 8, 1933.

Mrs. Franklin Roosevelt, accompanied by her son, Elliot, was scheduled to arrive at United Airport, Burbank, at approximately 2 this afternoon.

When Mrs. Roosevelt steps down from the plane it will complete the first transcontinental air trip ever made by a President's wife.

The Roosevelts were to fly from Tucson, Arizona, where Mrs. Roosevelt interrupted her journey yesterday to visit Mrs. Isabella Greenway. She was welcomed by her son.

No official reception will be held at the United Airport, Burbank, at the request of Mrs. Roosevelt.

"This is purely a non-political trip," she smiled.

While here, Mrs. Roosevelt will be the house guest of Mrs. Franklin K. Lane, Jr. of 179 South McCadden Place.

Eleanor Roosevelt, 1933

AVIATION

Army Planes to Note Air Races
1933

Three squadrons of Army and Navy Reserve and California National Guard combat airplanes will circle and land at Los Angeles municipal airport next Sunday afternoon at 2:30 o'clock as the first event of the colorful ground breaking ceremonies marking the beginning to the construction program for the National Air races which opens here July 1, according to word received today by Clifford W. Henderson, air race director, from army and navy officials.

Three Enter Race
1933

A racing team of May Haizlip, holder of the women's world speed record; J. R. Wedell, builder and pilot of racing planes, and Lee Gehlbach, racing pilot, today were officially entered in the principal high-speed events during the 1933 national air races at Los Angeles municipal airport, July 1, 2, 3, and 4.

L. A. to New York, 21 Hours, 45 Minutes

Half a day has been cut from the air schedule between New York and Los Angeles by 30 new transports going into service Sunday on the United Airlines.

The new schedule will be Los Angeles to New York, 21 hours and 45 minutes. New York to Los Angeles, 23 hours and 30 minutes.

Remote Control

Improvements in remote control equipment that will make possible intricate stunts by "robot" planes in midair were claimed today by Harry Reynolds and Howard G. Grove, Los Angeles electrical engineers.

Ox-Cart, Plane is Used by Man

Salt Lake City, June 6, 1933
From ox-cart to airplane is the span of travel modes used by Charles Shields, veteran Utah merchant and mining man. Shields came to Utah by ox-cart in 1870 and the other day, when starting a journey back to his native Ireland, left by airplane.

"It seems queer," Shields said, "that I can now fly to St. Louis in a few hours. My first trip from there to here took months."

Mollisons Again Postpone Flight

Dateline: London, June 6, 1933
James and Amy Johnson Mollison again postponed the take-off of their London-New York flight today because of unfavorable weather conditions over the Atlantic. They plan to fly to New York, then to Aleppo, Arabia and back to London.

Distance Flight Planned by Rossi

Dateline: Paris, June 7, 1933
A projected flight for the world's straight-line distance record was announced today by Maurice Rossi, French aviator.

He said he and Paul Codos would start from New York for the Persian gulf in the airplane Joseph Lebrix, in which Rossi and Lucien Boussoutrot recently tried a non-stop flight from France to Buenos Aires.

Dirigible, Thomas Scott Baldwin

AVIATION

Curtiss B-2 Formation over Atlantic City, New Jersey

Douglas World Cruiser

"Southern Cross"
Fokker F VII Trimotor Monoplane

AVIATION

Vin Flyer, Calbraith Perry Rodgers, 1911.

Eberhart SE-5E
U.S. Air Force Photo

AVIATION

De Havilland DH-4
U.S. Air Force

Drawings

The following drawings of aircraft were done by the author, Rosary, in late 1933, and they are included herein.

AVIATION

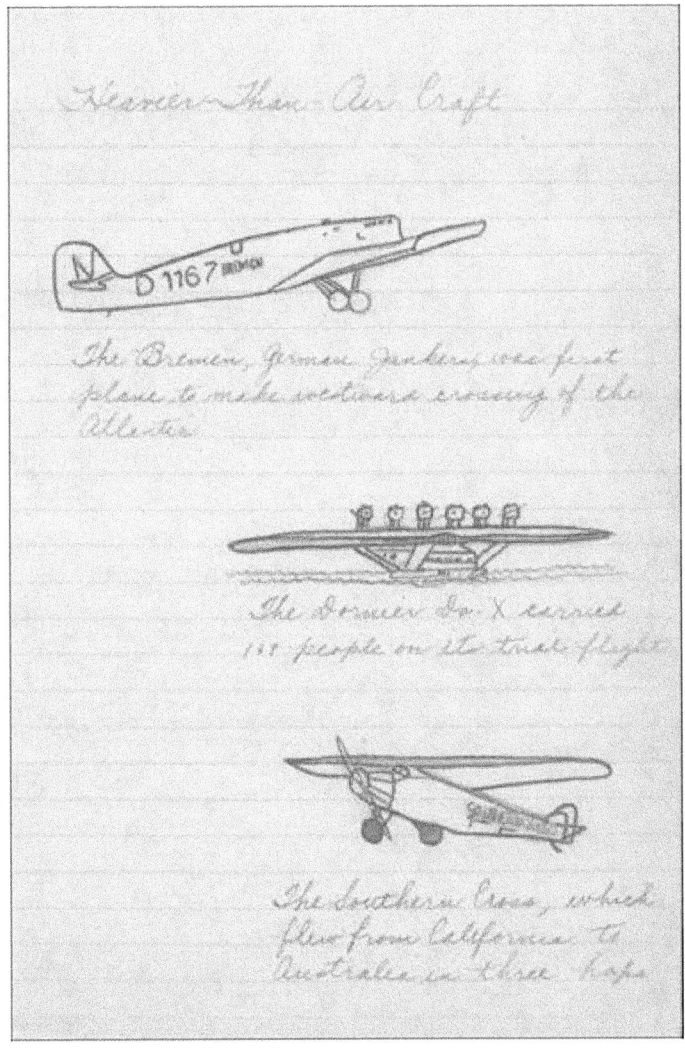

Heavier-Than-Air Craft

The Bremen, German Junkers, was first plane to make westward crossing of the Atlantic.

The Dornier Do. X carried 158 people on its trial flight.

The Southern Cross, which flew from California to Australia in three hops.

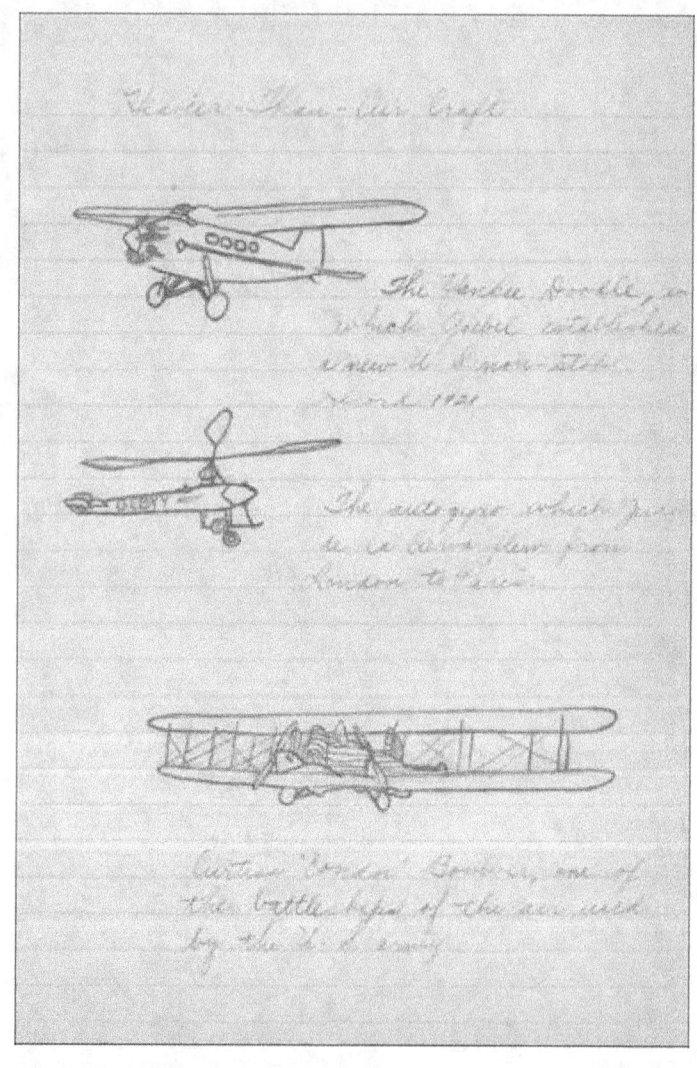

AVIATION

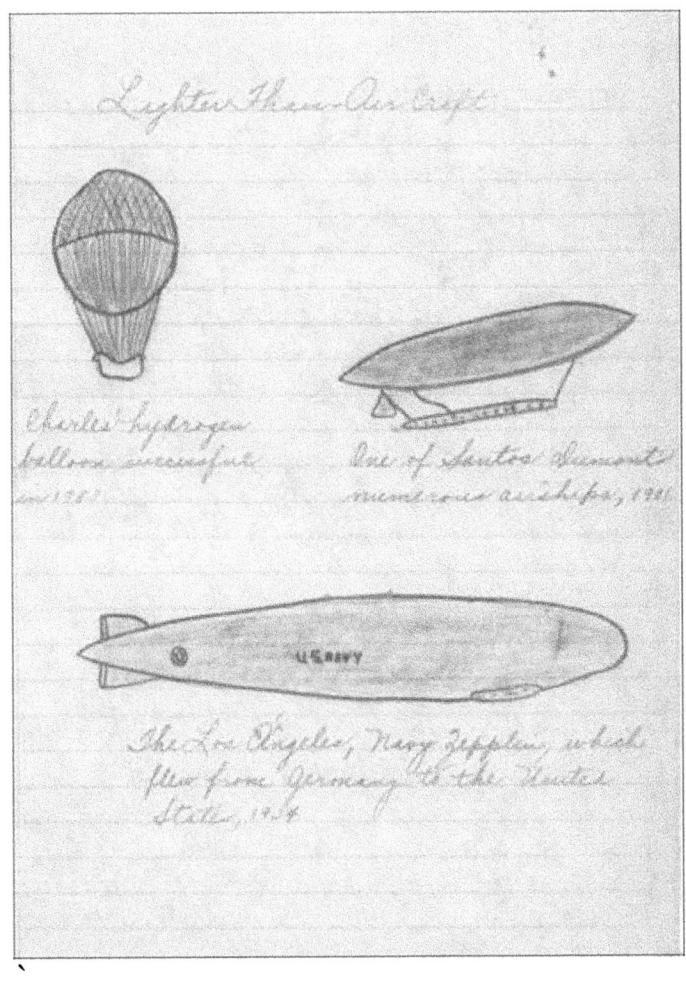

~end~

About the Author

1940

Rosary Spera was born in California in 1918. She was a first generation Italian, her parents having come to America from Sicily in the late 1800s. After graduating high school, she attended Woodbury College in Los Angeles, graduating in 1938. She married John Abrams in 1941, and they raised two daughters, Linda and Nancy. Rosary passed away in 2012, at the age of 93.

Foreword by Linda Pendleton, 2015.

Linda Pendleton has published a number of fiction and nonfiction books, poetry, and Comics. She also coauthored books with her late husband, Don Pendleton, known for his Executioner: Mack Bolan Series and numerous other books.

Linda makes her home in California with her Ragdoll cat, Max. She has a daughter and son, four grandchildren, four great-granddaughters, and another great-grandchild on the way. She enjoys family genealogy when not writing.

First Kindle Edition, 2015
First Print Edition 2020

Pendleton Artists
California

www.lindapendleton.com

www.ingramcontent.com/pod-product-compliance
Lightning Source LLC
Chambersburg PA
CBHW050252220526
45465CB00002B/644